Listening to Place

Listening to Place

Participatory Soundscape Planning for Communities

BRYAN C. PIJANOWSKI

Purdue University Press | West Lafayette, Indiana

Cataloging-in-Publication Data on file at the Library Congress.

978-1-62671-203-4 (paperback)
978-1-62671-204-1 (epdf)

Cover created by the author.

CONTENTS

Learning

Impacts of Noise on Humans and Wildlife

Sonic Spaces

Soundscape Participatory Workbook

Tools and Planning Resources

Action Plan

PREFACE

This *Listening to Place: Participatory Soundscape Planning for Communities* book is to be used by communities in the United States and possibly other places, to organize land-use planning within the context of the sonic spaces of built and natural areas locally. This book has several chapters:

- "Learning the Basics" helps planners understand what a soundscape is and what is noise.

- "Impacts of Noise on People and Wildlife" summarizes what is known about how sound affects people and wildlife.

- "Descriptions of Sonic Spaces" that may exist in communities including those that are natural, cultural and considered noise.

- "Soundscape Participatory Workbook" allows individual planners or community groups to work through major issues related to natural and cultural soundscapes and the occurrence of noise in their community.

- "Description of Tools and Planning Resources" at the national, state, and local levels for land-use planners.

- "Action Plan Worksheets" for advancing community objectives.

The importance of sound in land-use planning is often ignored, or, if it is part of the planning process, it is focused solely on the problem of noise. This book takes a different approach to addressing sound in community planning by considering the positive, as well as unwanted impacts, of sound, and in particular, soundscapes. As more than half of the world's population now lives in cities, a majority of society is exposed not only to the blanket of noise produced by the urban environment, but people in communities lack access to the calm, nurturing sounds of the natural world that have positive effects on well-being. This book helps to inform planners of the role that sound plays in affecting our overall mental and physical well-being but also directs attention to those sonic spaces in communities that could have economic and cultural impact. This participatory community planning book was generated as a partnership between members of Purdue's Center for Global Soundscapes and Purdue's Extension program in the College of Agriculture. The planning exercises that are contained here were first tested in workshop settings with urban planning offices of northwest Indiana communities. Funding for the project came from the Department of Forestry and Natural Resources endowments, the USDA McIntire-Stennis Program, the Great Lakes Restoration Initiative, Purdue's Graduate School, and the Indiana Department of Natural Resources Coastal Services Program.

1 LEARNING THE BASICS

What Is Sound?

Anything that moves makes a sound. Sound is a vibration or pressure wave that moves through space. It is a form of energy that causes groups of molecules to be bunched into high-pressure regions separated by molecules of lower pressure within an elastic medium, typically a liquid (e.g., water) or a gas (such as air). For a vibration or pressure wave to exist, an object has to physically generate waves that produce rapid changes in pressure that is passed from molecule to molecule across space creating a chain reaction until the energy dissipates. When these vibrations reach a human ear, they are collected by a membrane (the tympanum) and then passed to the hammer, anvil, and stirrup bones that amplify the vibration. This vibration is captured by the cochlea situated in the inner ear, which is lined with tiny hairs (cilia) that move and produce a nerve impulse to the brain. That is how we, mammals, hear. Other animals hear in various ways; we know that nearly all mammals, birds, reptiles, amphibians, and most insects hear or have ways to detect vibrations or sound pressure waves around them.

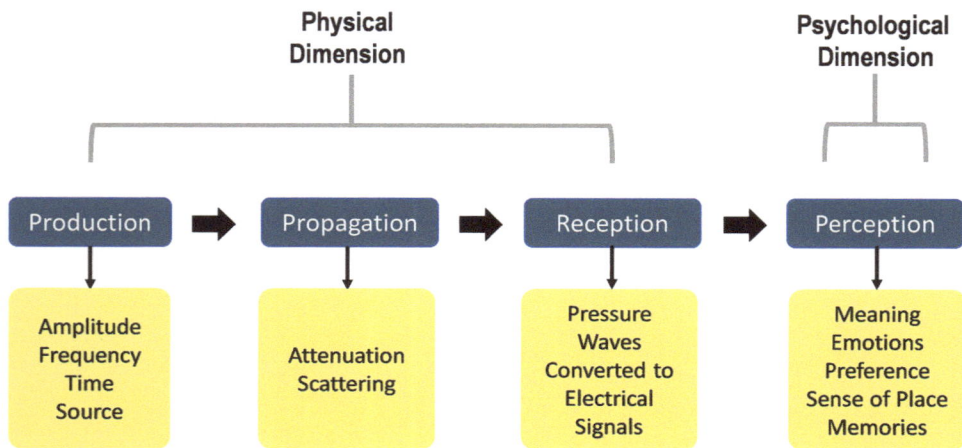

Figure 1. Standard sound model with the physical and psychological aspects of acoustic communication

To understand sound for the purposes of planning, it is helpful to understand the sound model (Figure 1) and its three main components. Sound **production** can be from *biological* (i.e., nonhuman living organisms), *geophysical* (e.g., thunder, rain, wind through trees, avalanches, ice breaking, and waves breaking on a shore), or those originating from *human activity* (e.g., machine noise, church bells, human voices, and music) sources. After sound is produced, it travels as a wave through a medium (air, water, or a solid substance such as the ground) to reach a target or receiver; only organisms can perceive and process information that sound might carry. Sound has three physical dimensions. The first is the amount of energy or its amplitude, this is measured as sound pressure level (SPL) often in units called decibels (dB). This dimension is also referred to as loudness. The second is frequency, or the number of cycles per second, the sound pressure wave travels through the medium. For humans, this is what we call pitch, but this is the dimension of sound that relates to whether the sounds that we perceive are high or low-pitched sounds. The third dimension is simply time. The length of the sound source is an important measure to scientists, engineers, and planners.

The physics of sound is studied by the science of acoustics that seek to understand how sound waves **propagate** through different media as well as the interaction between these waves and objects located across space (e.g., hills or a dense forest). Two physical changes can occur as sound propagates—attenuation (i.e., loss of energy as it travels through space and/or reduced by substances) and scattering (i.e., changes in the direction of the sound wave propagation).

However, in physiology and psychology, sound is traditionally **perceived** as a subject of perception by the brain. From this standpoint, if we ask the question "If a tree falls in the forest with no one to hear it fall, does it make a sound?" the psychological answer would be "No." However, from the perspective of acoustics, the answer to the previous question would be "yes," the tree will make a sound independently if it is heard by someone or not, as the pressure wave indeed existed in that space.

Humans and other animals perceive their environment through several cognitive layers. These include determining meaning (e.g., what is the sound source, how close, what direction), aligning sounds to emotions (e.g., sounds make me happy or fearful), determining preference (e.g., like or dislike), establishing a sense of place (e.g., associate sounds with place and time), and then cataloguing sonic experiences as memories. Common sounds within a community often form a sense of place which are often shared by everyone. If you live in the Midwest United States, for example, spring peepers, a small native frog of wetlands, signal the start of spring for many people. As people from your community enjoy city parks and other outdoor areas, are the sounds they hear contributing to the experience that these spaces are designed for? Are noise levels from the urban environment causing health problems?

What Are Soundscapes?

A **soundscape** can be defined as the collection of all sounds that occur in an area over a particular length of time. This term was promoted in the early 1970s by R. Murray Schafer,[1] a Canadian composer and environmentalist. Since then, it has been used by many disciplines to describe the association between landscapes and their sounds.[2]

Soundscape ecology[2] is the science that studies all sounds, those collectively as biophony (nonhuman, biological sounds), geophony (geophysical sounds), and/or anthrophony (human-produced sounds), that aid in our understanding of the ecology of landscapes (Figure 2). The use of the suffix -phony is used because scientists recognize

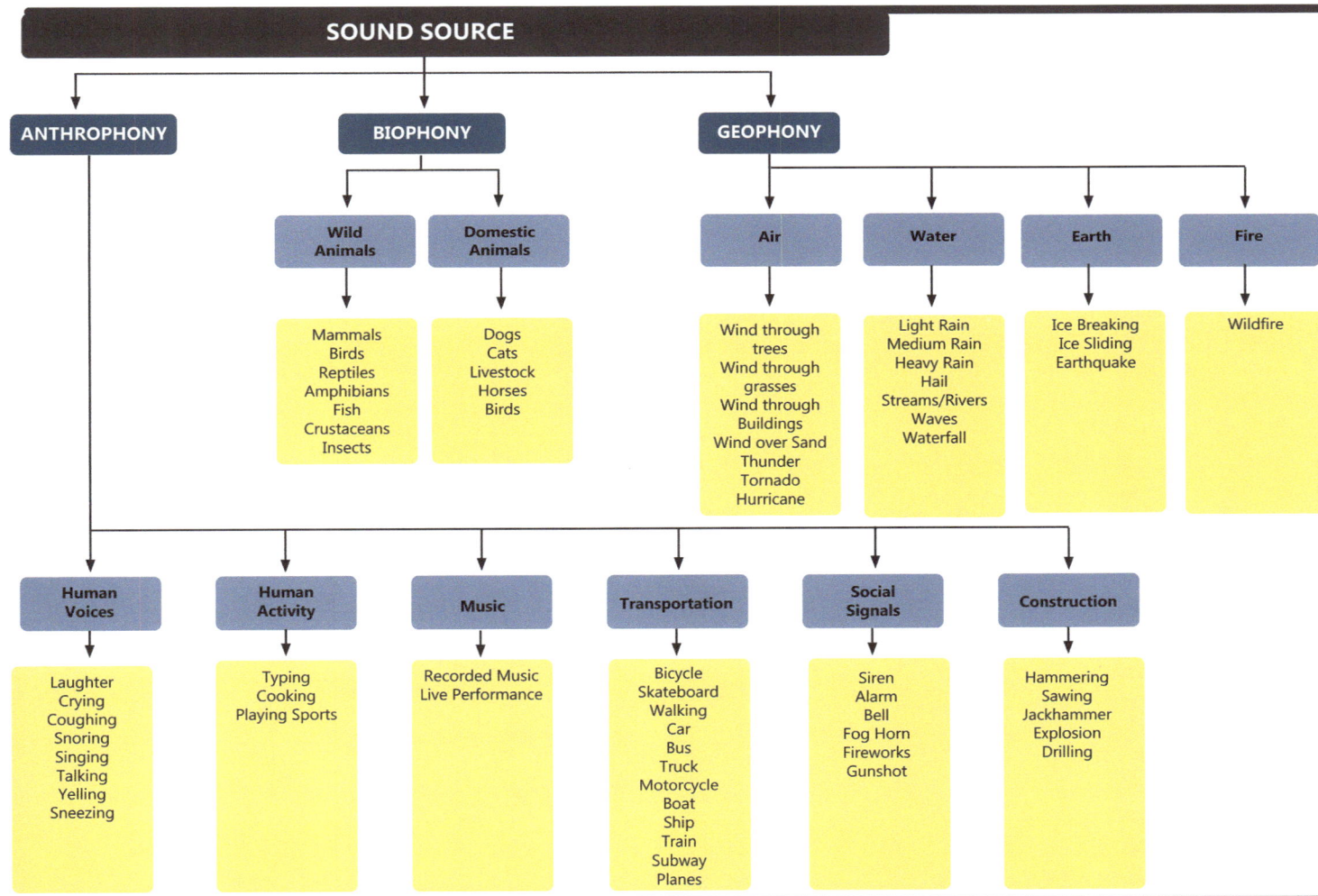

SOUND SOURCE

```
ANTHROPHONY        BIOPHONY                      GEOPHONY

              Wild        Domestic       Air        Water       Earth       Fire
             Animals      Animals

             Mammals        Dogs      Wind through   Light Rain   Ice Breaking   Wildfire
             Birds          Cats         trees       Medium Rain  Ice Sliding
             Reptiles       Livestock  Wind through   Heavy Rain   Earthquake
             Amphibians     Horses       grasses       Hail
             Fish           Birds      Wind through   Streams/Rivers
             Crustaceans               Buildings      Waves
             Insects                  Wind over Sand  Waterfall
                                        Thunder
                                        Tornado
                                        Hurricane
```

```
   Human          Human                                   Social
   Voices        Activity      Music      Transportation  Signals     Construction

  Laughter        Typing     Recorded Music   Bicycle       Siren      Hammering
  Crying          Cooking    Live Performance Skateboard    Alarm      Sawing
  Coughing        Playing Sports             Walking        Bell       Jackhammer
  Snoring                                     Car           Fog Horn   Explosion
  Singing                                     Bus           Fireworks  Drilling
  Talking                                     Truck         Gunshot
  Yelling                                     Motorcycle
  Sneezing                                    Boat
                                              Ship
                                              Train
                                              Subway
                                              Planes
```

Figure 2. Sound Taxonomy Diagram. As sounds are often related to the source producing it, the above illustration attempts to organize sound sources hierarchically across numerous groupings of sources that are useful for understanding ecosystem dynamics as well as for natural resource and land-use planning. This taxonomy does not necessarily have to be strictly followed and could be modified if the natural and cultural soundscapes and noise types significantly differ from those contained here.

that sounds from each of these groups is structured. For example, biological sounds have evolved by all animals living in the same place so that they all do not overlap. Applications of soundscape ecology include biodiversity assessments, planning for healthy living spaces, and increasing society's connectedness to nature. The field of soundscape ecology differs from century-old, traditional study in biology called **bio-acoustics.** Bioacoustics has traditionally focused on species specific sound production or perception behaviors of animals. Soundscape ecology, on the other hand, considers all sound sources and how they are perceived by people and animals.

The frequencies of common sound sources in the soundscape are illustrated in Figure 3 (below). Note that most of the biological sound sources—animals—exists in the 1,000 Hz to 8,000 Hz range, which is also the frequencies that are most sensitive for humans. Sounds from human activity tend to dominate in lower frequencies but, problematically, overlap and thus can mask the biological sounds. Sounds from the geophysical environment are also in the same frequencies that are sensitive to humans, but these also tend to have broadband structures (i.e., sounds that span large range of frequencies).

What Is the Distinction Between Sound and Noise?

Noise is a type of sound perceived by humans that causes an unpleasant experience; therefore, it is an unwanted sound. From this standpoint, people can subjectively evaluate the quality of the sonic environment regarding the level of background noise capable of masking important sounds.

Sounds that are considered to be noise can be rather subjective; to one person, the sound of a loud motorcycle might remind them of the pleasant experiences of riding such a vehicle during a sunny afternoon; to others, it might be an unwanted sound that unexpectedly interrupts their current activity. Some sounds are more universally accepted as noise, such as the sounds of a nearby highway or the constant sounds of jets flying overhead.

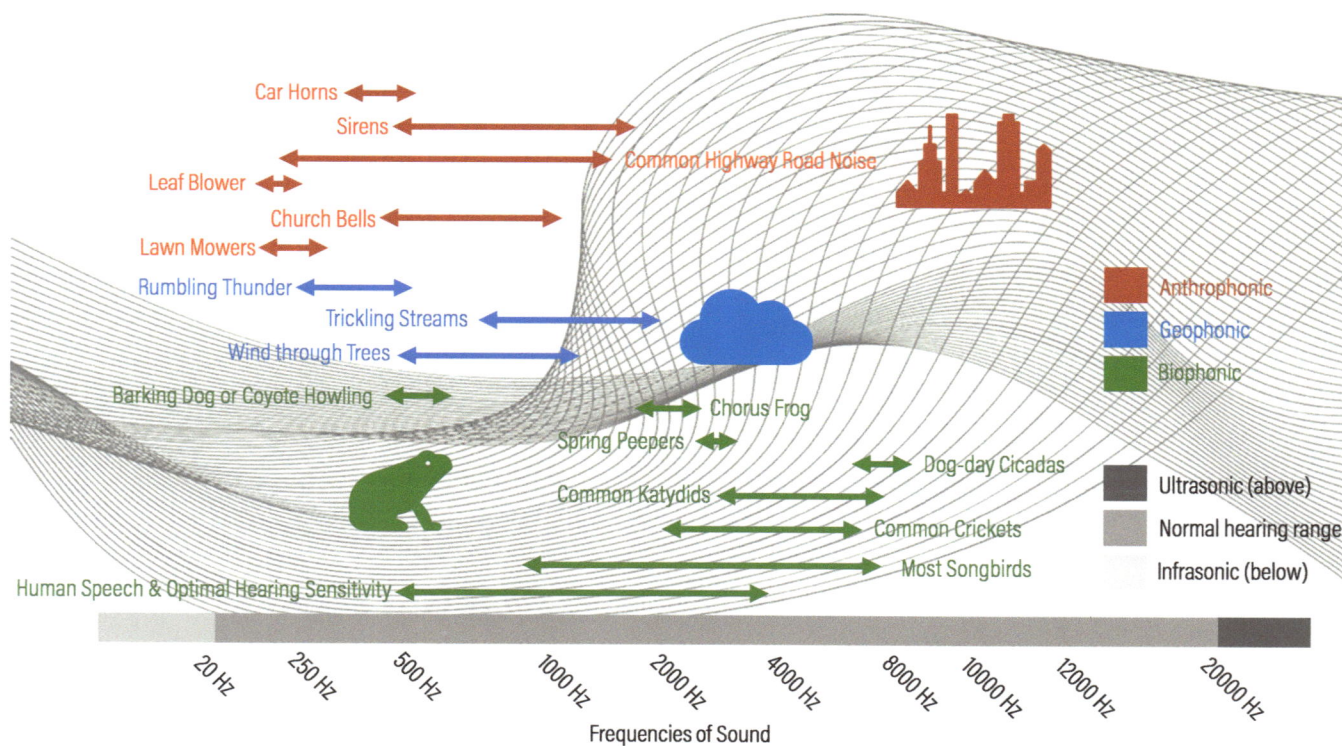

Figure 3. Common sound sources in a soundscape along with their frequencies

Noise and Cultural Sounds

There are many sounds produced by humans that may not be considered noise and are rather very important part of a community's identity. These might include the sounds of a church bell, the sounds produced by a foghorn or boat, or public musical performance.

Some human-produced sounds can be considered cultural and important in one community, and noise to yet others. For example, the sounds of trains in some communities may define that soundscape for members living there especially if there are historical or economic ties to that transportation system. Other communities may find those sounds to be unwanted and therefore noise.

Visualizing Sound

There are two common ways to visualize sound. The first is to plot amplitude over time (Figure 4A). This is called a **waveform**. The second is to plot frequency, time, and amplitude in a diagram called a **spectrogram** (Figure 4B). Note that for the spectrogram the x-axis (horizontal axis) is time in seconds; the y-axis (vertical axis) is frequency in Hz (Hertz or cycles per second), with amplitude often coded with brighter colors representing sound with more energy (i.e., loudness).

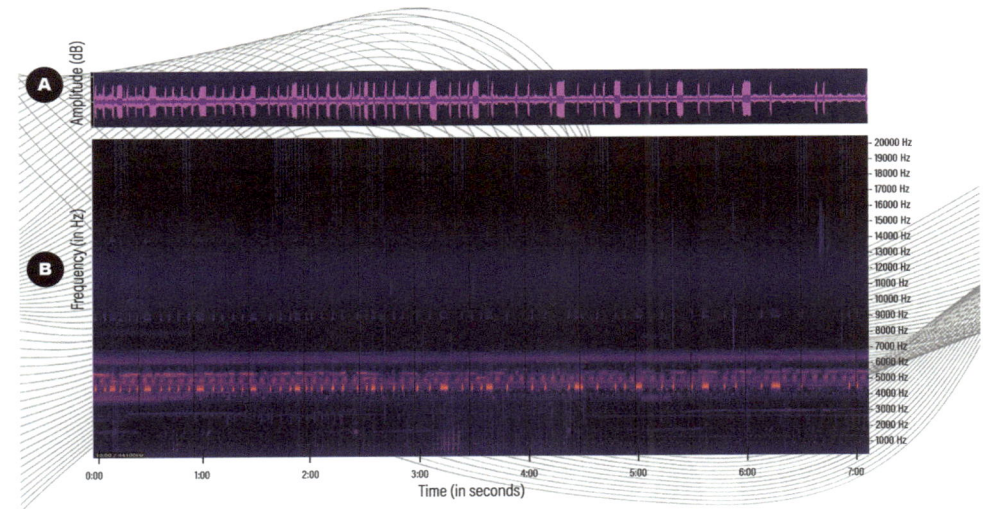

Figure 4. Visualizing sound as a (A) waveform for amplitude and (B) spectrogram for frequency, time and amplitude

Sound Intensity or Loudness

The measure of the intensity of sound is called its amplitude or sound pressure level (SPL). The most common way to express sound intensity is the decibel. The decibel scale represents the amplitude relative to the faintest sound a typical human can hear at a reference frequency of 4000 Hz. The decibel is calculated as ratio so that a 10 dB sound is one that is 10 times greater than the lower limit of the threshold of hearing. A 20 dB sound source is 100 times greater than the quietest sound a human can hear, 30 dB is 1000 times greater, and so on. The specific term loudness is used to refer to perceived amplitude by a person given the frequency of the sound and its amplitude. Loudness of the same sound will vary by individual. A list of common sounds and their amplitudes are listed in Table 1.

Table 1. Acoustic environment, average dBA levels and human response

Acoustic Environment	Average Decibels (dBA)	Human Response
Sounds barely detectable by person	0	—
Leaves rustling, soft music, whisper	30	Calm or sense of solitude
Average home noise	40	
Normal conversation, background noise	60	—
Office noise, inside care at 60 mph	70	Generates annoyance
Vacuum cleaner	75	—
Heavy traffic, window air conditioner, noisy restaurant, power lawn mover	80–89	Increased stress and hypertension
Subway, shouting	90–95	Long-term exposure creates hearing loss
Boom box, ATV, motorcycle	95–100	—
School dance from band	101–105	—
Chainsaw, leaf blower, snowmobile	100–110	
Sports crowd, rock concert	120	Short-term exposure creates hearing loss
Jet engine or power saw	120–130	
Shot gun, siren at 100 feet	140–150	Pain

Techniques for Measuring Soundscapes

In general, there are two ways that scientists and land-use planners have measured soundscapes. The first approach has been to do aural assessments by listening and noting the sound sources that occur at a place.[3,4] The second is to use sensors or acoustic instruments to quantify the soundscape.

Qualitative: This approach focusses on considerations of sonic spaces as a resource (e.g., are there spaces that are specific to culture and/or recreation?), what the sound sources are (e.g., make a list), and how the sounds should be managed (e.g., protected or abated). Often, soundwalks (e.g., strategic walking through areas with a map and noting sound sources) are performed during specific times of the day and/or times of the year. Another aspect of the qualitative assessment is to determine sound source preferences. How do citizens of communities react to sounds across a pleasantness (like to dislike), eventfulness (how often they occur), and familiarity (strong or weak)?[5]

Quantitative: This approach, used mostly by soundscape ecologists and bioacousticians, uses passive acoustic sensors (see Figure 5) that record sounds to a digital wave file in high fidelity, often at qualities that are used in professional movies and music recordings. This digital information is then analyzed by three kinds of acoustic tools. The first is to calculate diversity, complexity, and intensity values for all sound sources using acoustic indices. In general, the diversity and complexity of soundscapes are greatest in the morning and evening (e.g., dawn and dusk chorusing) and in natural areas rather than in human-dominated areas. The second set of tools involve the use of sound source artificial intelligence algorithms such as those found in the bird song identification app called Merlin,[6] that uses a massive library of bird calls to identify the calls of most of the world's bird calls. Similar tools are being developed by scientists to identify other sound sources.[7] Finally, acousticians often measure sounds as noise levels (see next chapter) and report these as averaged over time values.

Standards for Measuring Noise

In outdoor noise studies, acousticians will measure time-averaged sound pressure levels and report these as L_{eq}, or the equivalent continuous sound level, in decibels, for a specified time. In other words, it is a time averaged sound pressure level. These studies often employ the use of handheld instruments called sound level meters. Studies may report in dBA (also denoted as dB(A)), rather than dB, and so LA_{eq} is often used to denote these kinds of measurements. Other forms of L_{eq} include LeqD (for daytime), $L_{eq}N$ (for nighttime), $L_{eq}24$ (for the entire day), or $LA_{eq}18h$ (L_{eq} for dBA over the period of 06:00 to

Figure 5. A Purdue University scientist servicing a passive acoustic sensor in a forest

24:00 hours). L_{eq} can also be reported as statistical values, where a L_{eq} for sound levels exceeding a specified time are reported. For example, LA90 is the sound level that exceeds 90 percent of the measurement period. Finally, LDN is commonly reported for the 24-hour period that adjusts L_{eq} for the night using a 10 dB penalty for nighttime (usually between 10pm and 7am) measurements as humans are more sensitive to noise at night.

Another standard for measuring noise is to use models of sound propagation[8] to map noise profiles across space and time. There are numerous national, regional, and local companies that work with local communities and transportation organizations that can model outdoor noise in 2D and 3D environments.

2 IMPACTS OF SOUND ON PEOPLE AND ANIMALS

Impacts of Noise on Humans and Wildlife

Noise and Well-Being

As noise affects the health and behavior of humans and other animals,[1] it is considered a type of pollution, although it has historically been treated differently than other types of pollutants such as airborne chemicals, air particulate matter, or electromagnetic waves, among others. A summary of the ill-health and positive effects of different kinds of sounds with key references from the scientific literature that reviews the known research are in Table 2. Note that these health problems often co-occur and are thus not independent of one another.[2]

Noise can be measured with high precision with powerful instruments; however, it gains more meaning if we couple this measurement with human perception and a physiological sensitivity to these sounds. Every person has her/his own perception and sensitivity about noise, making thresholds variable among people. Studies have also focused on noise impacts on children, which suggest multiple negative impacts on physical and mental well-being.[3]

It is well known that people who are more sensitive to noise show an increase in heart rate and blood pressure than people who are more noise tolerant. The effect is more pronounced in some populations that are more highly vulnerable to any kind of stress, such as adults with hypertension, cardiovascular disease, or diabetes.

Other effects of noise on people result in what some psychologists call "learned helplessness," which is a situation when adults and children continue to fail at a task due to the overt presence of environmental stimuli that result in a repeated lack of responsiveness.[4] For example, when children from schools near noise sources (such as trains and highways) were asked to solve a puzzle within four minutes, 15 percent of them gave up before the time limit, while children from schools with quiet surroundings gave up at a much lower rate of 2 percent. This motivational reduction has been increasingly reported by teachers in schools with noisy environments.

Table 2. Effects of sound on human well-being

Human Health	Sound Sources	Key Citation
High blood pressure	Noise	Münzel, T., Gori, T., Babisch, W., and Basner, M. (2014). "Cardiovascular effects of environmental noise exposure" in *European Heart Journal*, *35*(13), 829–836.
Sleep disruption	Noise	Zaharna, M., and Guilleminault, C. (2010). "Sleep, Noise and Health" in *Noise and Health 12*(47), 64–69.
Lack of ability to focus	Noise	Lu, Z. L., and Dosher, B. A. (1998). "External Noise Distinguishes Attention Mechanisms" in *Vision Research 38*(9), 1183–1198.
Cardiovascular disease	Noise	Babisch, W. (2000). "Traffic Noise and Cardiovascular Disease: Epidemiological Review and Synthesis" in *Noise and Health 2*(8), 9–32.
Stress	Noise	Westman, J. C., and Walters, J. R. (1981). "Noise and Stress: A Comprehensive Approach" in *Environmental Health Perspectives 41*, 291–309.
Hearing loss	Noise	Natarajan, N., Batts, S., & Stankovic, K. M. (2023). "Noise-Induced Hearing Loss" in *Journal of Clinical Medicine 12*(6), 2,347.
Intensify problems associated with Alzheimer's	Noise	Cui, B., and Li, K. (2013). "Chronic Noise Exposure and Alzheimer Disease: Is There an Etiological Association?" in *Medical Hypotheses 81*(4), 623–626.
Dementia	Noise	Carey, I. M., Anderson, H. R., Atkinson, R. W., Beevers, S. D., Cook, D. G., Strachan, D. P., and Kelly, F. J. (2018). "Are Noise and Air Pollution Related to the Incidence of Dementia? A Cohort Study in London, England" in *BMJ Open 8*(9), e022404.
Calmness and Happiness	Natural sounds	Song, I., Baek, K., Kim, C., and Song, C. (2023). "Effects of Nature Sounds on the Attention and Physiological and Psychological Relaxation: in *Urban Forestry & Urban Greening 86*, 127987.
Caring	Natural sounds	Fisher, J. A. (1999). "The Value of Natural Sounds: in *Journal of Aesthetic Education 33*(3), 26–42.
Stress relief and attentive restoration	Natural sounds	Buxton, R. T., Pearson, A. L., Allou, C., Fristrup, K., and Wittemyer, G. (2021). "A Synthesis of Health Benefits of Natural Sounds and Their Distribution in National Parks" in *Proceedings of the National Academy of Sciences 118*(14), e2013097118.
Improve cognitive functions	Natural sounds	Abbott, L., Newman, P., and Benfield, J. (2015). "The Influence of Natural Sounds on Attention Restoration" (Doctoral dissertation, Pennsylvania State University).

Sound Production and Hearing in Animals

There are certain sounds that, because of their origin or loudness, may produce discomfort for people, as well as for animals. Generally, these are human-made sounds such as those we can find in a city, especially those produced by traffic such as road noise, horns, sirens, etc.

In the natural world, many animals—such as birds, mammals, reptiles, amphibians, fish, and insects—hear best at the frequencies they produce and have varying sensitivities to sounds of other frequencies. All these groups of animals produce a high diversity of sounds that are influenced by several factors, including habitat, body size of the organism, and the medium in which the sounds are produced (e.g., water or air). Animals produce sounds for many reasons, such as finding mates, defending a territory, or for warning of predators. Some sounds from the natural environment (e.g., the sounds produced by rivers) are known to be used for orientation.

How Animals Produce and Sense Sound

There are a variety of animals that produce sounds and nearly all animals sense sounds in one form or another.[5] The list of important groups, with examples of sound-producing groups and/or species, include:

- mammals, such as squirrels, canines (e.g., dogs, wolves, coyotes), ungulates (e.g., deer, elk), and livestock (e.g., cows, sheep, pigs)

- nearly all birds, with many being iconic to places, such as the common loon

- amphibians (e.g., toads, frogs)

- and insects in purposeful ways; the most common are crickets, katydids, and cicadas

Some animals produce sounds from their activities; woodpeckers strike trees with their beaks, producing a pecking sound, and insects, which produce sounds from other activities. For example, sounds from wingbeats are often audible and detectable with acoustic sensors. These include wingbeat sounds from are flies, bees and wasps, mosquitoes, which tend to be species specific.

Because some forms of acoustic production travels best in certain kinds of landscapes, the sounds of birds in forests tend to be dominated by whistles, whereas sounds in open grassland areas tend to have a "buzzier" characteristic to them.

Finally, it should be noted that many aquatic animals, such as those living in lakes and ponds, produce and sense sound. This includes aquatic mammals, most fish, and many species of aquatic invertebrates.

Cultural Values and Soundscapes

Besides natural and human-made sounds and noise, there is another aspect of the sonic environment that deserves attention: the cultural value of specific sounds. Cultural sounds consist of those sound sources, which have certain social values related to traditions, contemporary culture, or habits. Sounds associated with these social values are cultural sounds.

Sense of Place

We create different types of relationships with the place we belong. For instance, we can connect with a place from a historical or familial perspective, such as where we were born or lived in a place for a period. Additionally, we can generate spiritual connections with a place that is emotional or intangible rather than generated. We can connect with a place from ideological relationships when moral and ethical bonds are developed, as when we follow or live according to certain philosophical guidelines, which could be religious or secular. We can also generate narrative relationships with a place through stories, myths, family histories, politics, and fiction. Each of these relationships between people and places forms a sense of place that makes a particular place unique.

Sounds are an important component of personal connection to place[6] because they influence emotional and aesthetic values we assign to spaces. Sounds help to generate affective psychological bonds between people, their culture, roots, and values. For example, the whistle of a train, the calls of street vendors, or children playing in parks are all sounds that people feel connected with and may define their belonging to a particular place. This point of view also considers that sounds form a "sense of belonging to us," if we consider that values of sounds may be shared by the community, family, or an individual.

While our environment changes with time, cultural soundscapes may change as well. For example, when we build new places or modify existing ones, we incorporate new sounds into the environment, changing those preexistent soundscapes that many have depended on psychologically. We might also shutter certain places such as churches, train stations, or schools that used to produce sounds important for members of a community. In this case, we might be losing cultural soundscapes that are impossible to restore.

Nature Connectedness

As people have become physically separated from natural settings, it has been argued by environmental psychologists that the level of nature connectedness—which is the extent to which individuals include nature as part of their representation of self[7] is declining across our society and also through time. It is argued here that by introducing more natural sounds exposure opportunities to members of urban, suburban, and even rural areas, members of communities could increase nature connectedness, which has been proven to increase mental well-being.[8]

Noise and Wildlife Behavior

Noise is generally a product of urban development and industrialization. In addition, it modifies the acoustic environment of natural areas (on land or even in aquatic environments). Even pristine natural areas do not escape noise pollution. One study across 22 US national parks showed that noise was, on average, audible more than 28 percent of the time.

There have been a lot of studies by scientists that have demonstrated that urban noise alters the behavior of bird calls.[9,10] For example, it has been found that some bird species sing at higher frequencies in urban environments to transmit their signals above the lower frequencies of road noise, some species sing with greater intensity in areas that have a high volume of road noise, and that other bird species will even sing at night rather than during the day due to the fact that the night has less urban sounds.[11,12,13] Finally, for aquatic systems, scientists have found that the sounds of boat motor noise occur in most frequencies that fish use to communicate to conspecifics (i.e., members of the same species) and can affect important life functions like spawning success.[14]

3 SONIC SPACES

Sonic Spaces

Scholars in natural resource management have often considered the sonic space in both air and water to be a resource just like trees, water, and soils.[1] Indeed, these scholars have argued that natural sounds and quiet conditions are common pool resources; these are resources that can be utilized by people and animals. When an individual introduces other sounds (e.g., loud noise) into these spaces, then they tend to have a subtractive effect on those that can benefit from these sonic conditions.

Sonic Spaces

We argue here that there are many places in communities where natural or cultural sounds and/or quiet spaces need to be protected so that their value is not reduced.[2] We label, and describe briefly, sonic spaces that communities might consider either protecting and/or regulating through specific ordinances (see Chapter 5):

Natural Sound Spaces: For many communities, parks, and other recreational places are important amenities that are managed by parks associations. These could include areas that contain natural vegetation, streams, trails (walking or biking), some perhaps augmented with nature centers that educate the public about the natural world around them. Consider educating the public about the different kinds of natural sounds that exist in these spaces, their natural history, and how they change through time (e.g., what does spring sound like and why?). Building an acoustic connection of place to the natural settings will create a more nature-connected community. Parks should also be designed so that noise from highways and other high transportation centers (e.g., airports, bus stations) are not permeating these natural sonic spaces. Other areas that fall into these sonic space categories include natural preserves, state, local and national parks, forests, wildlife refuges, wild and scenic rivers, and wilderness areas. Most highway sounds can be heard over a mile away although many factors such as speed of vehicles, atmospheric conditions, and pavement surface conditions influence this factor.[3] Areas that have been identified as contemplative areas (e.g., for creating a feeling of solitude with natural sounds being dominated) should be especially protected and well-planned to avoid disruption by unwanted sounds of noise.

Recreational Sonic Spaces: Spaces for hunting, fishing, bird watching, and other recreational activities. These spaces could overlap or be the same as those for natural sound spaces. Here, though, the sound of recreational activity (e.g., sounds of fishing) is likely to be an important aspect of these activities and so they need to be protected (e.g., areas should be protected from unwanted noises). Although there is not a lot of research on this topic, the effects of loud water vehicles have been shown to change the behavior of aquatic animals such as fish, waterfowl, and crustaceans, thus possibly introducing conflicts in recreational uses (e.g., fishing versus boating) in lake ecosystems.[4,5]

Cultural Sonic Spaces: There are many sounds in communities that have cultural value. These include those that are symbolic sounds (e.g., church bells or the evening prayer chants of Muslims) or those that distinctly define spaces. For the latter, the sounds of streetcars or trolleys, which have been preserved historically, could be sounds that are valued by communities and have a desire to be protected. The sounds of trains (e.g., rumbling along tracks, train whistles) can also be a common sound source that defines a community's soundscape. Trains are often on a strict schedule and so many members of a community that have lived in a place for a long period of time have come to expect certain trains at a specific time of the day. The sounds of combines or livestock are important cultural soundscapes of rural communities. In the United States, the sounds of high school football games on Friday evenings are a soundscape that is often experienced either at the sporting event or by nearby communities. Finally, cultural soundscapes can also include music, the sounds of fishing wharfs in coastal communities, or the sounds of street vendors hawking their wares.[6] Indeed, certain kinds of music often adds to the "vibe" of any downtown area.

High-Noise Sonic Spaces: Many communities have high-level noise spaces that are unavoidable. These include communities that have highway intersections, national and international airports, and loud commercial operations. Land-use planners should consider avoiding placing the following new developments in these high-level noise areas: schools, nursing homes, hospitals, nature areas, protected areas, and recreational areas.

Nature-Based Solutions

Soundscape planning is part of a larger, popular and ever-growing paradigm of planning called nature-based solutions, which focusses on sustainability of urban systems.[7] Also designated as NBS (nature-based solutions), this approach to planning is "inspired by nature, use nature and/or are supported by nature."[8] This means that planning that includes principles of green infrastructure, which promotes the development and integration of natural areas for the purposes of stormwater management, water purification, and increasing recreational spaces, is likely to align well with planning to improve natural and cultural sonic spaces in a community.

Landscape-Soundscape Relationships for Planning

Ecologists that study plant and animal relationships at landscape scales have found that there are important relationships between habitat features and animal biodiversity. These relationships include the following generalizations:

Size of Habitat Boosts Animal Biodiversity: Overall, larger patches of natural habitat increase the diversity of the animal species that live there. This is because there are some species that dominate in edge environments (e.g., the edge of a forest that abuts a crop) while others prefer core habitats. For example, the oven-bird, pileated woodpeckers, yellow-throated vireos, and the American redstart, are common songbirds that only live deep in forests. Birds that prefer natural habitat edges include the American robin, yellow warbler, and the wood thrush. The size of wetlands also increases the diversity of songbirds and waterfowl that utilize these kinds of natural areas.

Habitat Structural Complexity Increases Animal Diversity: Recent research by scientists shows that the structural complexity of a habitat in both the vertical (i.e., up and down) and horizontal (i.e., across the landscape) increases animal biodiversity, and thus increases the complexity and diversity of the soundscape. Ensuring that old-growth trees are maintained as well as areas of dead trees provides the increased diversity and complexity of natural area habitats.

Increase Connectivity Increases Animal Biodiversity: Increasing the connectivity of natural areas increases the ability of animals to move through the landscape and also supports large populations of the same species. Landscape connectivity is the amount that a landscape facilitates the movement of organisms through space. Habitat patches, for example, should either be connected by physical corridors or be close enough for most animals to traverse across gaps. Roads and other barriers need to be managed so that animals can move from patch to patch. Natural soundscapes are facilitated by landscape connectivity.

Natural Habitats Support Different Soundscapes: Whereas forests support certain kinds of songbirds, mammals and insects, wetlands support waterfowl, amphibians, and certain kinds of insects. Forests, in general, will contain bird calls that span whistles to melodies and insects that typically are pulsating (e.g., cicadas) whereas wetlands will support many rhythmic animals such as amphibians and stridulating insects.

4 SOUNDSCAPE PARTICIPATORY WORKSHEETS

Soundscape Participatory Workbook

This chapter helps community planners to think through all the issues related to planning for natural and cultural soundscapes and to potentially identify areas of concern where noise is present. It describes a sequence of exercises framed by motivating questions and suggestions for how these activities can be accomplished using participatory or coproduced approaches.

Exercises in this and the following chapters include:

Exercise 1: Sonic Brainstorming, which describes an activity to list as many sounds as possible that occur in your community.

Exercise 2: Soundscape Mapping, where the location of noise, natural and culture soundscapes and noise zones are identified.

Exercise 3: Sonic Timelines, which help to determine when certain kinds of sounds occur so that they can be properly addressed.

Exercise 4: Ear-Appeal Assessment associates psychological impacts to sound sources that occur in your community.

Exercise 5: Who Should Care? is a participatory guide to help generate a list of organizational resources (groups, tools) that can be employed for community soundscape planning.

Exercise 6: Regulations and Laws focus on gathering a list of local, state, and federal laws that regulate noise in communities based on sound sources (e.g., airports, highways, railways).

Exercise 7: Planning for Healthy Sonic Spaces is designed to provide space to organize action plans for noise, natural sounds, and cultural sounds.

Exercise 1: Sonic Brainstorming

Motivating Question of this Activity: What are the sounds of my community?

How to Do This: You might consider having everyone write down, on colored sticky notes (one color each for biological, geophysical, and anthropogenic) sound sources in their community. Also on that sticky note, include scores for noise levels (1=low, 5=very loud) and arrows up (for keep) and arrows down (for remove). These can then be summarized in the table below for Exercise #1 and then saved for Exercise #4. Discussions with the community and local government planning staff could focus on those sounds that are keynote sounds.

Soundscape Participatory Workbook

EXERCISE 1: SONIC BRAINSTORMING

What are the sounds of your community?

Cultural Sounds	Natural Sounds	Noise Sources and Levels from 1 (quiet) to 5 (very loud)

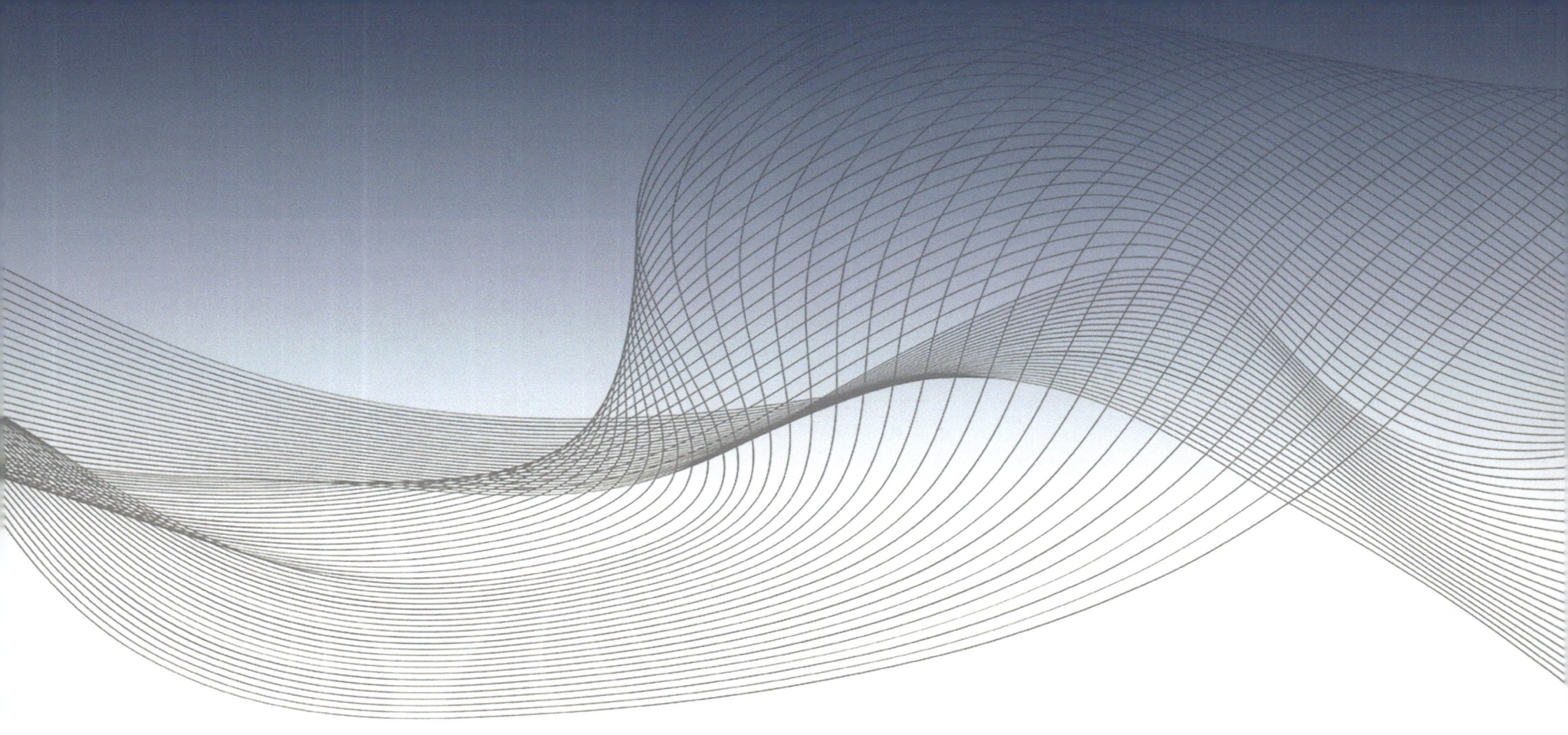

Exercise 2: Soundscape Mapping

Motivating Question of this Activity: Where do these sounds occur and are there soundscape and noise "zones" that our community should consider?

How to Do This: A large-format printer could be used to print a map of the community that is then placed on a table. Participants can then use colored pens to label areas with those sounds in each of the three major categories. Please include cultural and natural sounds, as well as noise sources and relative levels. These boundaries along with labels could then be added as a GIS layer to any online spatial visualization tools that a community uses to communicate with the public. Additional activities could include having volunteers from local nature or recreational groups go out and record the sounds and create a map of local soundscapes. Alternatively, small groups of local citizens led by planning office staff could host soundwalks where the group walks strategically through areas documenting the sounds of the places.

EXERCISE 2: SOUNDSCAPE MAPPING

Where do these sounds occur and are there sonic spaces and/or noise zones that our community should consider?

Please include: cultural and natural sounds, as well as noise sources and relative levels.

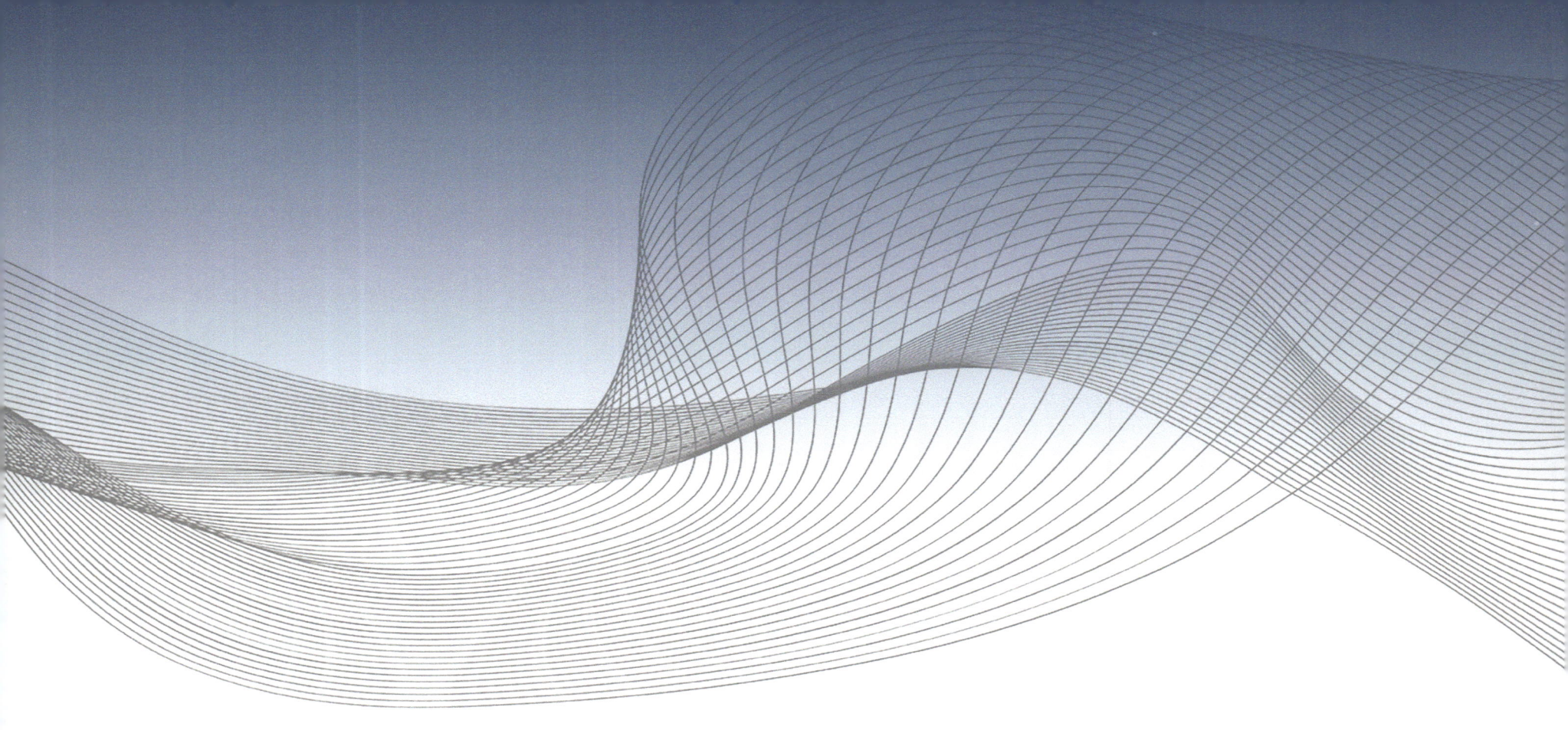

Exercise 3: Sonic Timelines

Motivating Question for this Activity: How do the soundscapes of my community change over time.

How to Do This: Time horizons could include those that are hourly, day of week, seasonal, or are centered on certain events (festivals). Locations that were mapped in Exercise 2 could be considered for this temporal assessment. Keep in mind that noise levels during the day and those at night are very important to consider as these have health implications. You might also think about time-based restrictions. For example, implementing quiet hours for nature parks (e.g., early in the morning when people are birding) or having events on viz., Earth Day (April 22nd of every year), that encourage people to go out and listen to nature.

EXERCISE 3: SONIC TIMELINES

List sounds and noise and describe how they change over time. Time horizons should be hourly, day of week, or seasons/events.

When is highway traffic the greatest? When do community members experience natural sounds? Are there times of the year when certain sounds (e.g., gunshot from hunters) are most common?

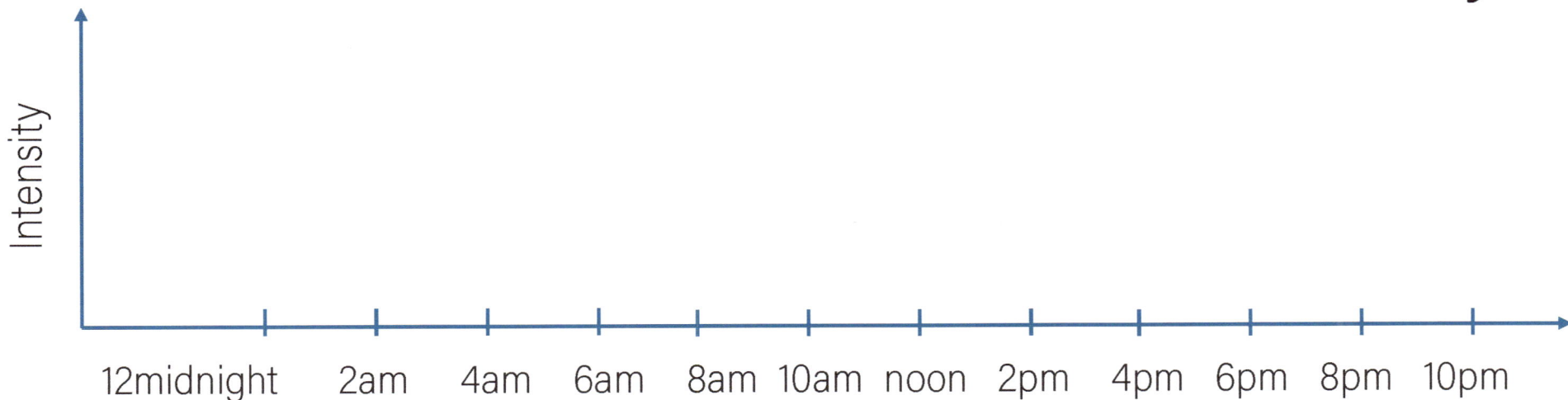

Hourly

Intensity

12midnight 2am 4am 6am 8am 10am noon 2pm 4pm 6pm 8pm 10pm

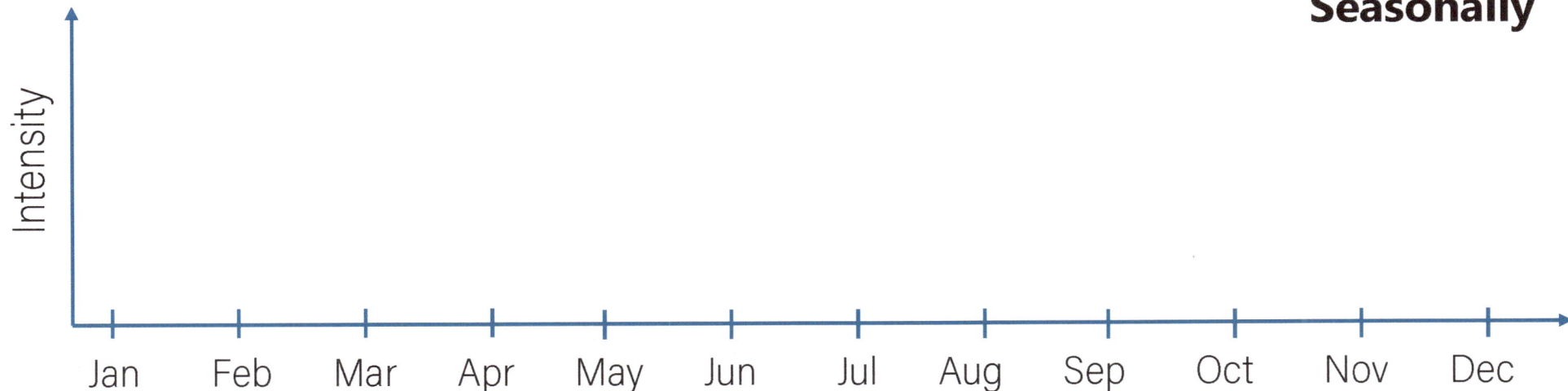

Seasonally

Intensity

Jan Feb Mar Apr May Jun Jul Aug Sep Oct Nov Dec

Exercise 4: Ear-Appeal Assessment

Motivating Question for this Activity: How do members of your community feel about sounds that occur in your locality and should they stay or be removed?

How to Do This: Use the table below to have each participant fill these out or do this as a group activity. Please consider each sound below and indicate your assessment of by circling both an emoji and thumbs up (stay) or down (remove). Refer to the taxonomy in Chapter 1.

How do members of your community feel about sounds that occur in your community and should they stay or be removed?

Please consider each sound, below and indicate your assessment of by circling both an emoji and thumbs up (stay) or down (remove). Refer to the taxonomy on the opposite page.

ANTHROPONY

Category	Sound	Stay/Remove	Emoji
Transportation	Bicycle	👍👎	😀🙂😐🙁☹️
	Skateboard	👍👎	😀🙂😐🙁☹️
	Walking	👍👎	😀🙂😐🙁☹️
	Car	👍👎	😀🙂😐🙁☹️
	Bus	👍👎	😀🙂😐🙁☹️
	Truck	👍👎	😀🙂😐🙁☹️
	Motorcycle	👍👎	😀🙂😐🙁☹️
	Boat	👍👎	😀🙂😐🙁☹️
	Ship	👍👎	😀🙂😐🙁☹️
	Train	👍👎	😀🙂😐🙁☹️
	Subway	👍👎	😀🙂😐🙁☹️
	Planes	👍👎	😀🙂😐🙁☹️
Social Signals	Siren	👍👎	😀🙂😐🙁☹️
	Alarm	👍👎	😀🙂😐🙁☹️
	Bell	👍👎	😀🙂😐🙁☹️
	Fog Horn	👍👎	😀🙂😐🙁☹️
	Fireworks	👍👎	😀🙂😐🙁☹️
	Gunshot	👍👎	😀🙂😐🙁☹️
Construction	Hammering	👍👎	😀🙂😐🙁☹️
	Sawing	👍👎	😀🙂😐🙁☹️
	Jackhammer	👍👎	😀🙂😐🙁☹️
	Explosion	👍👎	😀🙂😐🙁☹️
	Drilling	👍👎	😀🙂😐🙁☹️

BIOPHONY

Category	Sound	Stay/Remove	Emoji
Wild Animals	Mammals	👍👎	😀🙂😐🙁☹️
	Birds	👍👎	😀🙂😐🙁☹️
	Reptiles	👍👎	😀🙂😐🙁☹️
	Amphibians	👍👎	😀🙂😐🙁☹️
	Fish	👍👎	😀🙂😐🙁☹️
	Crustaceans	👍👎	😀🙂😐🙁☹️
	Insects	👍👎	😀🙂😐🙁☹️
Domestic Animals	Dogs	👍👎	😀🙂😐🙁☹️
	Cats	👍👎	😀🙂😐🙁☹️
	Livestock	👍👎	😀🙂😐🙁☹️
	Horses	👍👎	😀🙂😐🙁☹️
	Birds	👍👎	😀🙂😐🙁☹️
Human Voices	Laughter	👍👎	😀🙂😐🙁☹️
	Crying	👍👎	😀🙂😐🙁☹️
	Coughing	👍👎	😀🙂😐🙁☹️
	Snoring	👍👎	😀🙂😐🙁☹️
	Singing	👍👎	😀🙂😐🙁☹️
	Talking	👍👎	😀🙂😐🙁☹️
	Yelling	👍👎	😀🙂😐🙁☹️
	Sneezing	👍👎	😀🙂😐🙁☹️
Human Activity	Typing	👍👎	😀🙂😐🙁☹️
	Cooking	👍👎	😀🙂😐🙁☹️
	Playing Sports	👍👎	😀🙂😐🙁☹️
Music	Recorded Music	👍👎	😀🙂😐🙁☹️
	Live Music	👍👎	😀🙂😐🙁☹️

GEOPHONY

Category	Sound	Stay/Remove	Emoji
Air	Wind through trees	👍👎	😀🙂😐🙁☹️
	Wind through grasses	👍👎	😀🙂😐🙁☹️
	Wind through buildings	👍👎	😀🙂😐🙁☹️
	Wind over sand	👍👎	😀🙂😐🙁☹️
	Thunder	👍👎	😀🙂😐🙁☹️
	Tornado	👍👎	😀🙂😐🙁☹️
	Hurricane	👍👎	😀🙂😐🙁☹️
Water	Light rain	👍👎	😀🙂😐🙁☹️
	Medium rain	👍👎	😀🙂😐🙁☹️
	Heavy rain	👍👎	😀🙂😐🙁☹️
	Hail	👍👎	😀🙂😐🙁☹️
	Streams/Rivers	👍👎	😀🙂😐🙁☹️
	Waves	👍👎	😀🙂😐🙁☹️
	Waterfall	👍👎	😀🙂😐🙁☹️
Earth	Ice Breaking	👍👎	😀🙂😐🙁☹️
	Ice Sliding	👍👎	😀🙂😐🙁☹️
	Earthquake	👍👎	😀🙂😐🙁☹️
Fire	Wildfire	👍👎	😀🙂😐🙁☹️

Exercise 5: Who Should Care?

Motivating Question for this Activity: What people or groups should be consulted about as planning activities move forward?

How to Do This: For this brainstorming exercise, work as a team to list some possible organizational resources that can be employed for community soundscape planning. These might include individuals, groups, or tools that would be helpful.

EXERCISE 5: WHO SHOULD CARE?

List some possible organizational resources that can be employed for community landscape planning.

These might include individuals, groups, or tools that would be helpful.

Contact Name	Contact Info	Resource Category (i.e. Company, Person or Tool)

5 TOOLS AND PLANNING RESOURCES

Noise and Land-Use Planning

Noise has historically been regulated in a variety of ways, including at the federal, state, and local levels.

Federal Guidance

With the aim of controlling and reducing noise pollution, national, state, and local government produce regulations such as Acts and Ordinances. The Noise Pollution and Abatement Act of 1972 (also referred to as the Noise Control Act) is a public law that was enacted in October of 1972. This act promotes an environment free of noise for all US citizens. The act serves to, a) guide effective coordination of federal research and activities in noise control, b) authorize the establishment of Federal noise emission standards for products distributed in commerce, and c) inform about noise emission and noise reduction characteristics of such products. In 1982, however, federal funding for this act ceased and the regulation of noise was moved to local and state government control.

Two ways that the federal government still considers noise impacts on the local environment is through transportation-related regulations such as airport, railway, and highway noise, and by requiring projects funded by federal agencies such as the Housing and Urban Development to meet noise standards or mitigate the impact of environmental noise.

Federal Aviation Administration (FAA)

The FAA has been working since 1981 to reduce the number of people impacted by airport noise. It established a program to assist airport operators in reducing the impacts of airports on communities. One way the program works is that airports are granted funds to buy adjacent property and prevent development. Another way is to improve existing structures so that they permit less noise to enter. The third way airports act to reduce the impact of aircraft noise is to map the sound impact on the land, identify land uses that are less vulnerable to noise, and route air traffic and runway approaches over those land uses. All of these activities are carried out under the FAA's "Part 150 Program."[1]

Federal Highway Administration (FHWA)

The FHWA studies the noise impact of future highways as well as mitigation measures such as noise barriers and berms.[2]

Federal Railroad Administration (FRA)

The FRA can approve "quiet zones" in communities where adequate rail crossing safety improvements have been made. Locomotive engineers are required to sound their horn in advance of all crossings, except in these special zones. Municipalities typically partner with railroads, state departments of transportation, and the FRA to plan, improve, and designate segments of urban rail corridor as quiet zones. Quiet zones do not affect the noise generated by trains as they pass along the rails, they only reduce the use of train horns. Use of train horns is described in detail in the Train Horn Rule (49 CFR Part 222, effective 08/17/2006).[3]

Federal Department of Housing and Urban Development (HUD)

HUD[4] provides guidelines for all funded and insured projects that must meet minimum requirements for a suitable living environment. It has established that noise levels above 65dB DNL (day-night average) are not acceptable. Normally unacceptable noise levels (but potentially acceptable if attenuation improvements are made) are 65–75dB DNL. Noise levels above 75 dB DNL are considered unacceptable and will require attenuation using changes in construction approaches. These noise levels are measured from all sources and are taken outside the structure. In general, the guidance is there should be a review for any projects that are within 1,000 feet of a major roadway, 3,000 feet from a railroad, and 15 miles in the vicinity of a new airfield.

State Guidance

Some states offer additional guidance on noise ordinances although most focus on occupational noise exposure in the work environments. A few states (e.g., Alaksa, Arizona, California, New York, and Hawaii) suggest noise control ordinances and penalties that local communities can adopt.

Local Guidance

Local Noise Ordinance Pertaining to "Nuisances"

In Indiana, each township passes its own set of noise regulations. Generally, ordinances have a standard structure. For example, they dedicate a section to definitions of the terminology used among the documents that may vary between ordinances, such as noise, sounds, decibels, sound level, motor vehicles, property boundaries, etc. Regulations also apply to activities that generate noise and are prohibited or regulated (in some cases, forbidden during certain times of day), such as the use of vehicles without a muffler, the use of a horn or signaling for an unreasonable period of time, operating radio, TVs, instruments, or loudspeakers that produce sound louder than the necessary volume to be heard, street sales, loading operations, construction activities, etc. Exemptions to the prohibition of using noisy devices or equipment are established, such as the use of sirens or emergency-related sounds, gardening machinery, and celebrations. Some or all of these kinds of sounds can be defined as nuisances in local laws. Township noise ordinances could benefit from evaluating spatial and temporal variables that reflect current noise problems within township boundaries.

A clearinghouse of noise pollution ordinances is found at the Noise Pollution Clearinghouse web site (nonoise.org). Here, resources are posted on noise regulations in the United States along with educational resources, information about conferences and meetings, engineering firms that specialize in noise, soundproofing resources, and a list of common noise sources to consider during the development or revision of noise ordinances are posted.

Some ordinances have special sections dedicated to animals and vehicles that may be very common noise sources. For instance, Michigan City and Chesterton, both towns in Indiana, regulate owning animals that cause noise disturbance such as barking dogs, birds that sing loudly, or roosters. Ordinances in places such as Porter, Indiana, and the city of Indianapolis, explicitly mention the noise levels allowed. For example, these municipalities regulate the decibel level (dB) of motor vehicles at given speeds and distances. Enforcement and penalties are common to ordinances, should rely on scientific measures of sound to ensure the ordinance is applied without bias, and measurements are defensible when challenged. Local police are responsible for enforcing noise regulations related to public nuisances. Fines or other penalties can be applied to residents, depending on the jurisdiction and details of the complaint.

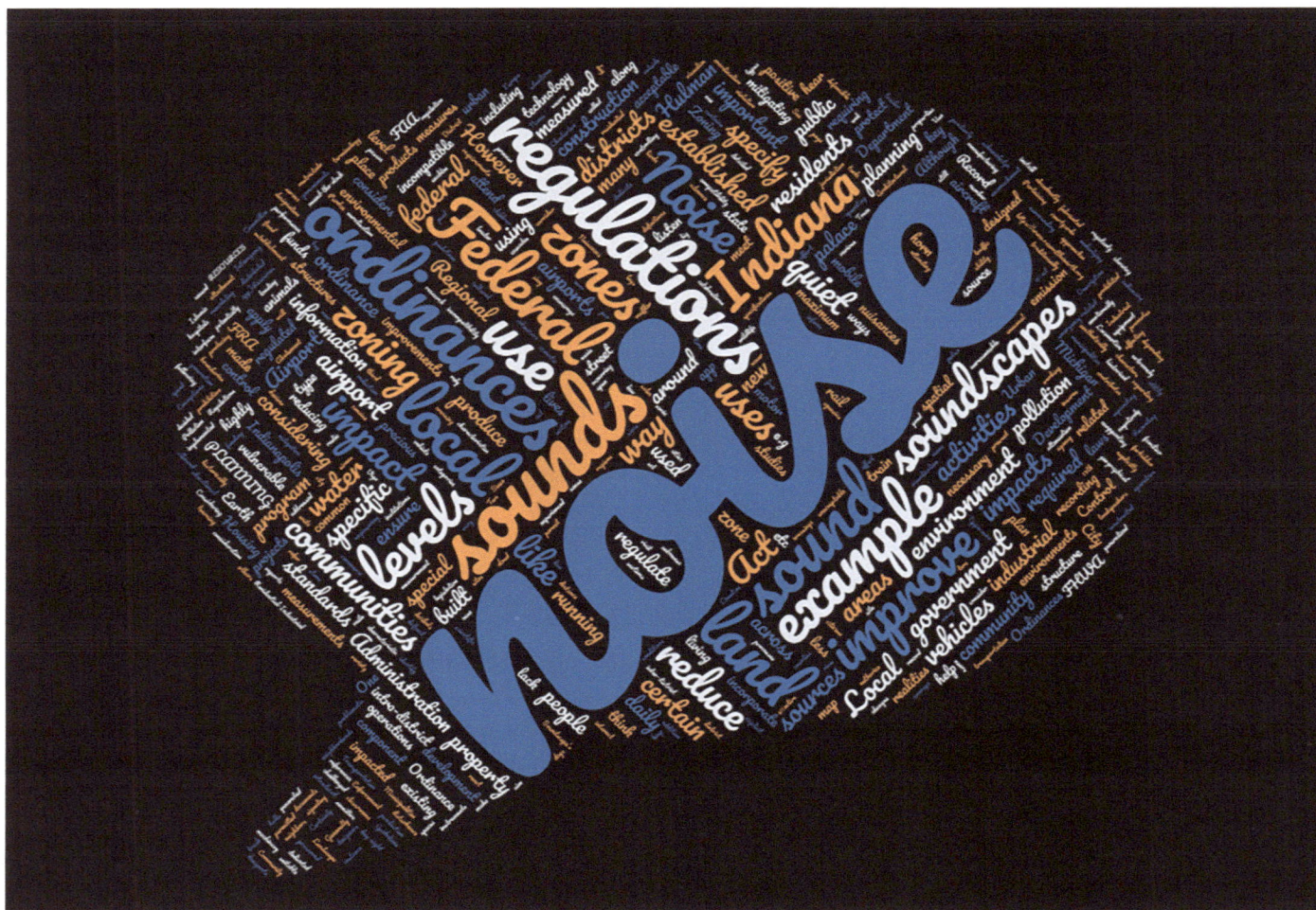

Many local government organizations in the United States and Canada support citizen-complaint or comment hotlines by dialing 311. These 311 nonemergency numbers provide citizens with access to services such as acquainting information about programs, procedures, events, reporting potholes, abandoned vehicles, as well as noise complaints. Many local government organizations also participate in the International Noise Awareness Day the last day of April of every year. This day is set aside for educating the public about the harmful effects of noise on people.

Local Zoning Ordinances

Local ordinances are designed to manage general noise situations of local residents. However, the current ordinances and laws may lack specifications that manage current noise problems in urban environments. For example, zoning ordinances in Indiana largely follow traditionally established patterns of separating incompatible land uses. Some zoning ordinances are not specific enough with situations between (residential-industrial) districts and within (industrial processes) districts. Residential-industrial districts are present in Indiana and occur when a tractor trailer manufacturer or gravel mine are located within view from residential homes. Industrial processes include noise produced by industry, such as manufacturing equipment, loud air conditioners, etc. Some Indiana ordinances, such as Tippecanoe County's Unified Zoning Ordinance, specify permitted noise levels within each district, defining the origination and units of the noise measurement. See the following chart that shows performance standards for noise generation in industrial and residential zoning districts.

Table 3 defines performance standards for categories A (adjacent rural and residential), B (adjacent commercial), and C (boundary line zones), and maximum permitted sound level. To consider intradistrict noise impacts, property line measurements would be used in analysis that measures the noise impact between two properties within the same zoning assignment.

Landscape regulations are also found in the zoning ordinance. Trees and other greenery can diffuse sound and provide an aesthetic barrier to sources of noise. In some zoning districts, these barriers may be required as a mitigating feature.

Table 3. Noise Limits per Tippecanoe County's Unified Zoning Ordinance (4-10-6-b-1) as Related to the US HUD Acceptability Standards

Performance Standard Category	Description	Maximum Permitted Sound Pressure Level (dBA)	Point of Measurement
A	Adjacent rural and residential	55 dBA daytime (7am–9pm) and 45 dBA nighttime (9pm–7am)	On adjacent rural and residential land use
B	Adjacent commercial	60 dBA	On adjacent commercial land uses
C	Boundary line zones	65 dBA	Across I1, I2, and I3 zone boundary lines

Exercise 6: Regulations and Laws

Motivating Question for this Activity: What are the regulations and laws in your community that help you to plan?

How to Do This: Use the table below to list the name of existing local ordinances, community resources (e.g., 311 call centers), or any land-use zoning laws that guide planning that could impact soundscapes across your community. Organize these as a separate list on the land-use planning web site for others to learn about these resources.

EXERCISE 6: REGULATIONS and LAWS

What are the regulations and laws in your community that help you to plan?

Name	Description

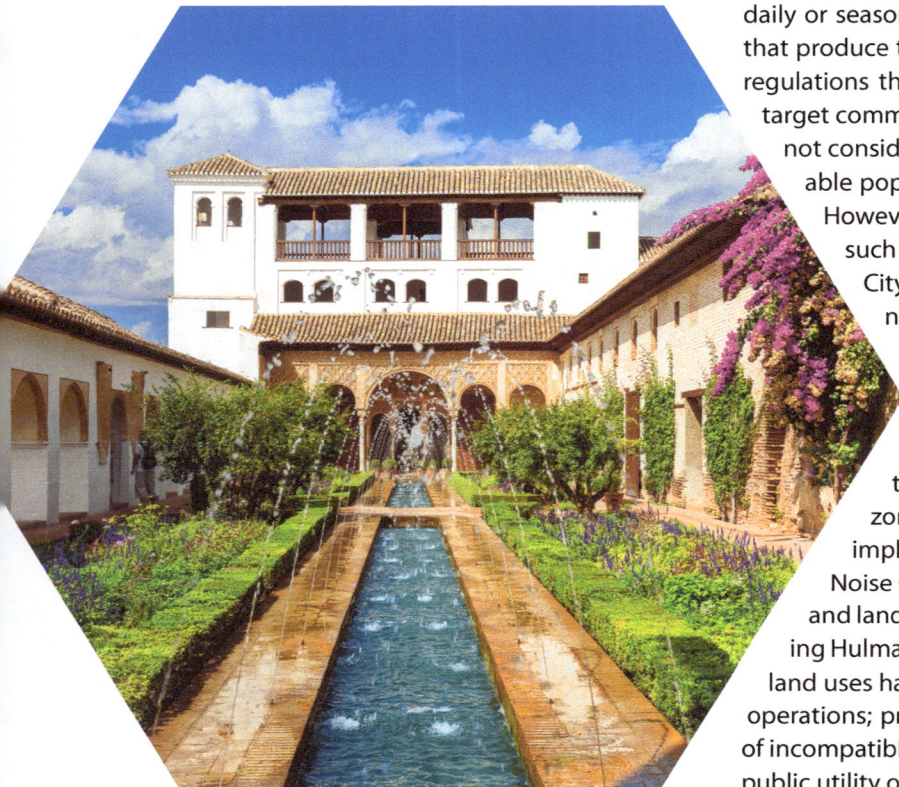

daily or seasonal) soundscape patterns and the sources that produce these sounds. In this way, we can improve regulations that affect specific geographical regions in target communities. For example, noise regulations do not consider the effects of noise pollution on vulnerable populations, such as the elderly and children. However, there are positive examples in cities such as Indianapolis, Fort Wayne, and Michigan City, that specific quiet zones around which noise production is restricted. These zones include hospitals, educational institutions, and churches. Another type of zone that can be used is an overlay zone. These apply specific regulations in addition to the existing regulations of the "underlying" zone. Vigo County, Indiana, for instance, has implemented the "Hulman Regional Airport Noise Overlay District" to "regulate development and land use within noise sensitive areas surrounding Hulman Regional Airport; encourage the types of land uses having maximum compatibility with aircraft operations; protect the airport from the encroachment of incompatible land uses; and, protect and promote the public utility of Hulman Regional Airport."

Communities can improve their sonic characters through mitigation but also by improving the quality and type of natural sounds and soundscapes. These sonic characteristics reduce noise for residents and community members while providing a habitat for wildlife.

For example, a new technology called auralisation (digital visualization using sounds) enables us to listen to virtual acoustic environments that have only existed in the past, that are about to be built, or that are fabricated. Such tools could be utilized in community planning in an effort to preserve beneficial soundscapes or to assist in examining problematic soundscapes.

Architects and environmental engineers use this approach in remodeling or for the construction of new concert halls and classrooms. Why not use this technology in open spaces to improve our daily lives? Certainly, we can interact more positively with our soundscapes by enhancing those sounds that are positive for us and reduce those that negatively affect us.

Novel Local Approaches to Soundscape Planning—Examples and Ideas

First, we can incorporate a spatial component into noise regulations. For example, if we know that noise affects learning we can specify an ordinance requiring that the construction of schools occur in quiet areas, as well as reduce current noise levels around school zones. Secondly, communities should implement a noise-monitoring program.

Although some local governments regularly measure noise levels in many areas within their communities, it is necessary to have more detailed information about the spatial (e.g., across neighborhoods) and temporal (e.g.,

Natural and Cultural Soundscapes

Natural and cultural soundscapes are an important dimension of planning. We might think of a historical example, the Alhambra in Granada, Spain. This is an Islamic palace built in the thirteenth century by Muslims who colonized Europe between 711–1492 AD. As they used to inhabit the desert back home, water was highly precious to them. The calmness that the sound of running water produces was surely familiar to them and highly precious. Because these colonizers were now living in a "European oasis," they decided to incorporate the sound of running water into the palace. Thus, they constructed a system of fountains and cascades in the courtyards and rooms within the palace with water running and falling through structures, producing extremely relaxing sounds. This could be thought of as one example of ancient planning to improve soundscapes of the built environment designed to affect the daily lives of residents.

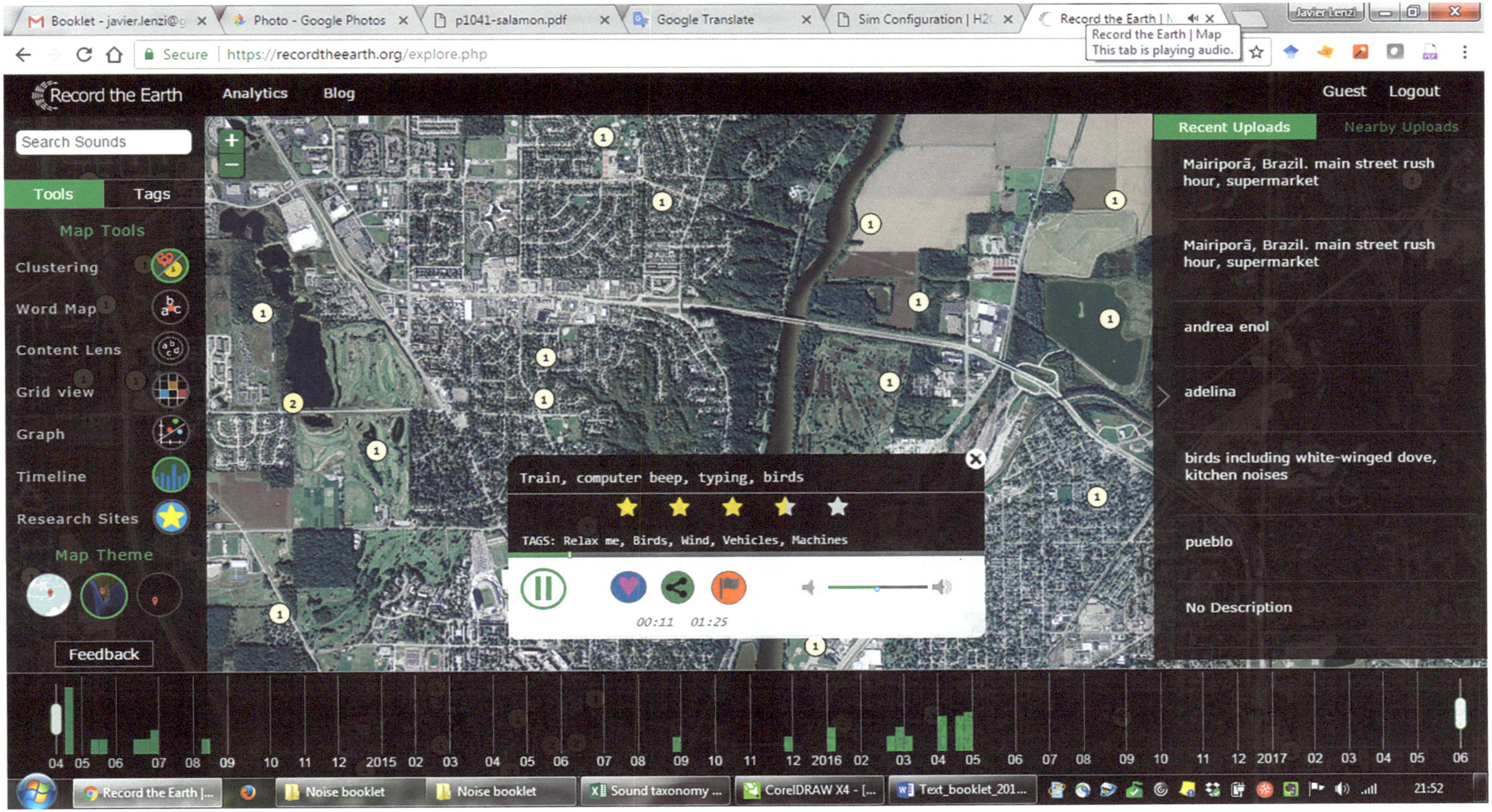
Booklet - javier.lenzi@g
Photo - Google Photos
p1041-salamon.pdf
Google Translate
Sim Configuration | H2
Record the Earth | M
Record the Earth | Map
This tab is playing audio.
Secure
https://recordtheearth.org/explore.php
Record the Earth
Analytics
Blog
Guest
Logout
Search Sounds
Tools
Tags
Map Tools
Clustering
Word Map
Content Lens
Grid view
Graph
Timeline
Research Sites
Map Theme
Feedback
Recent Uploads
Nearby Uploads
Mairiporã, Brazil. main street rush hour, supermarket
Mairiporã, Brazil. main street rush hour, supermarket
andrea enol
adelina
birds including white-winged dove, kitchen noises
pueblo
No Description
Train, computer beep, typing, birds
TAGS: Relax me, Birds, Wind, Vehicles, Machines
00:11
01:25
04 05 06 07 08 09 10 11 12 2015 02 03 04 05 06 07 08 09 10 11 12 2016 02 03 04 05 06 07 08 09 10 11 12 2017 02 03 04 05 06
Record the Earth |...
Noise booklet
Noise booklet
Sound taxonomy ...
CorelDRAW X4 - [...
Text_booklet_201...
21:52

6 DEVELOPING AN ACTION PLAN

Action Plan

Action Plan

This chapter is designed to provide you with some simple space to organize action plans for noise, natural sounds, and cultural sounds. The following three sections are part of the Noise and Soundscape Planning Action Plan. Use the information from the above worksheets and associated activities to summarize plans for noise, natural soundscape management, and cultural soundscape management. Ideally, this action plan should serve as a roadmap for implementing some broad management strategies that relate to soundscapes in your community. These include setting goals/objectives, identifying resources for action items and then a simple description of how the community moves from the goals to reality.

EXERCISE 7: PLANNING FOR HEALTHY SONIC SPACES
Toward an action plan for your community
This section of the workbook is designed to provide you with some simple space to organize action plans for noise, natural sounds, and cultural sounds.

Noise Monitoring and Abatement

Objectives What would you like to accomplish? Provide a list or narrative of what you would like to accomplish in this broad area of sonic space management.	
Planning Resources What tools, organizations, technologies, etc. are available to you to move toward your objectives?	
Action Plan State how you would move toward your objectives listing short-term (30 days to 1 year) and long-term outcomes (1 to 20 years) with strategies for attaining these within the time frame you describe.	

EXERCISE 7: PLANNING FOR HEALTHY SONIC SPACES
Toward an action plan for your community

This section of the workbook is designed to provide you with some simple space to organize action plans for noise, natural sounds, and cultural sounds.

Natural Sounds

Objectives What would you like to accomplish? Provide a list or narrative of what you would like to accomplish in this broad area of sonic space management.	
Planning Resources What tools, organizations, technologies, etc. are available to you to move toward your objectives?	
Action Plan State how you would move toward your objectives listing short-term (30 days to 1 year) and long-term outcomes (1 to 20 years) with strategies for attaining these within the time frame you describe.	

EXERCISE 7: PLANNING FOR HEALTHY SONIC SPACES

Toward an action plan for your community

This section of the workbook is designed to provide you with some simple space to organize action plans for noise, natural sounds, and cultural sounds.

Cultural Sounds

Objectives What would you like to accomplish? Provide a list or narrative of what you would like to accomplish in this broad area of sonic space management.	
Planning Resources What tools, organizations, technologies, etc. are available to you to move toward your objectives?	
Action Plan State how you would move toward your objectives listing short-term (30 days to 1 year) and long-term outcomes (1 to 20 years) with strategies for attaining these within the time frame you describe.	

GLOSSARY

Amplitude: The magnitude of pressure changes, which can be measured in various ways.

Compression and rarefaction: Compression refers to the areas of increased molecule density because of the high pressure. On the other hand, there are spaces where we can find a low density of molecules. This is called rarefaction.

dBA: This measurement is called A-weighted decibels (dBA). This unit has been adopted for measurements of environmental noise for example, to measure roadway noise, construction noise, and railroad noise, as well as to assess potential hearing damage as a result of noise. The A-weighting is an adjustment of the standard dB measurement based on the sensitivity of a normal person's hearing to sound pressure levels across frequencies.

Decibel(dB): A unit expressing the amplitude of a sound.

Frequency: Is defined as the number of occurrences of a repeating event per unit time. Technically is the speed of changes in pressure, or the speed of the vibrations. We can measure frequency by counting the number of wave cycles or periods per second (Hertz).

Keynote: sounds that are characterize a place, often due to the presence of an animal that makes a unique sound (e.g., common loon in northern lakes of the US).

Noise: Sound that impedes communication or interferes with life functions.

Sound: A pressure wave generated by vibrating objects that travels through a medium (e.g., air or water).

Sound pressure: Sound pressure is the pressure measured caused by the sound wave relative to the surrounding air pressure. Loud sounds produce sound waves with relatively large sound pressures, while quiet sounds produce sound waves with relatively small sound pressures. Like other kinds of pressure, it is commonly measured in units of Pascals (Pa).

Sound pressure level: Sound pressure level is a measurement of the effective pressure of a sound relative to a reference pressure that is typically the threshold of human hearing (2×10^{-5} Pa). Because human ears are sensitive to a very wide range of sounds, sound pressure level uses a logarithmic scale (a shorter scale) to represent the sound pressure.

Wavelength: The waves that produce sound can be represented in space as the figure below. The distance between two peaks in the plot is called wavelength.

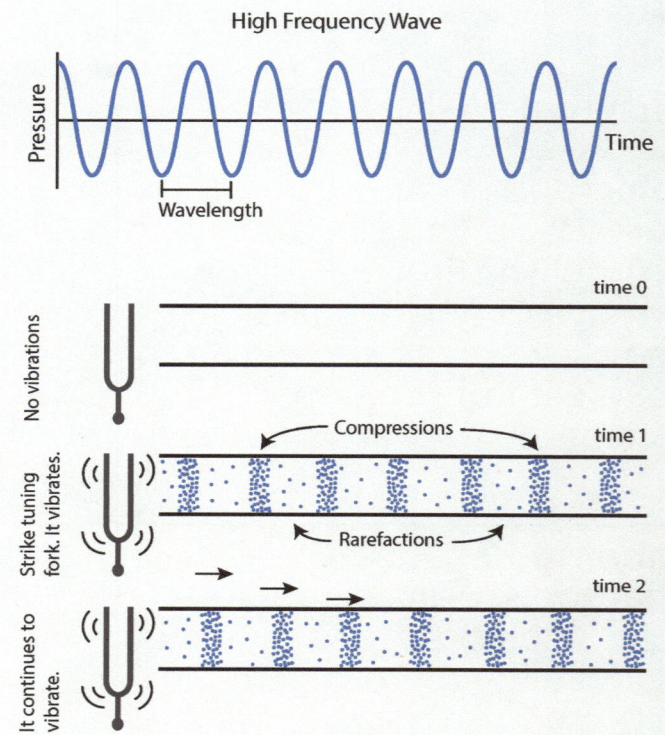

NOTES

CHAPTER 1

1. Schafer, R. M. (1993). *The Soundscape: Our Sonic Environment and the Tuning of the World.* Simon and Schuster.

2. Pijanowski, B. C., Villanueva-Rivera, L. J., Dumyahn, S. L., Farina, A., Krause, B. L., Napoletano, B. M., and Pieretti, N. (2011). "Soundscape Ecology: The Science of Sound in the Landscape" in *BioScience* 61(3), 203–216.

3. Brown, A. L. (2010). "Soundscapes and Environmental Noise Management" in *Noise Control Engineering Journal* 58(5): 493–500.

4. Pijanowski, B. C. (2024). *Principles of Soundscape Ecology: Discovering Our Sonic World.* University of Chicago Press, Chapter 13, "Urban Soundscapes."

5. Axelsson et al. (2010). "A Principal Components Model of Soundscape Perception" in *Journal of the Acoustical Society of America* 128 (5): 2836–2846.

6. Merlin, a bird song identification mobile app for Android and iOS from the Cornell Laboratory for Ornithology, Ithaca, New York.

7. Pijanowski, B. C., Fuenzalida, F. R., Banerjee, S., Minghim, R., Lima, S. L., Bowers-Sword, R., and Savage, D. (2024). "Soundscape Analytics: A New Frontier of Knowledge Discovery in Soundscape Data" in *Current Landscape Ecology Reports* 9(4), 88–107.

8. Alam, P., Ahmad, K., Afsar, S. S., and Akhtar, N. (2020). "Noise Monitoring, Mapping, and Modelling Studies—A Review" in *Journal of Ecological Engineering* 21(4).

CHAPTER 2

1. Francis, C. D., Ortega, C. P., and Cruz, A. (2009). "Noise Pollution Changes Avian Communities and Species Interactions" in *Current Biology* 19(16), 1415–1419.

2. Stansfeld, S. A., and Matheson, M. P. (2003). "Noise Pollution: Non-Auditory Effects on Health" in *British Medical Bulletin* 68(1), 243–257.

3. Evans, G. W., and Lepore, S. J. (1993). "Nonauditory Effects of Noise on Children: A Critical Review" in *Children's Environments*, 31–51.

4. Dohmen, M., Braat-Eggen, E., Kemperman, A., and Hornikx, M. (2022). "The Effects of Noise on Cognitive Performance and Helplessness in Childhood: A Review" in *International Journal of Environmental Research and Public Health* 20(1), 288.

5. Pijanowski, B. C., F. R. Fuenzalida, J., Kang, C. Weaver, A. Franics, M. McPherson, M. Heinz, J. Post, O. E. Rhodes, C. Mediastika, T. G. Sors, T. H. Park and S. Grimes (2025). "A Sonic One Health Framework: Characterizing Health of the Environment, Animals, and Humans Using a Soundscape Perspective" in *BioScience*.

6. Tuan, Y. F. (1979). "Space and Place: Humanistic Perspective" in *Philosophy in Geography*, 387–427. Dordrecht: Springer Netherlands.

7. Mayer, F. Stephan, et al. (2009). "Why Is Nature Beneficial? The Role of Connectedness to Nature" in *Environment and Behavior* 41.5: 607–643.

8. Pritchard, A., Richardson, M., Sheffield, D., and McEwan, K. (2020). "The Relationship between Nature Connectedness and Eudaimonic Well-Being: A Meta-Analysis" in *Journal of Happiness Studies* 21, 1145–1167.

9. Fuller, R. A., Warren, P. H., and Gaston, K. J. (2007). "Daytime Noise Predicts Nocturnal Singing in Urban Robins" in *Biology Letters* 3(4), 368–370.

10. Engel, M. S., Young, R. J., Davies, W. J., Waddington, D., and Wood, M. D. (2024). "A Systematic Review of Anthropogenic Noise Impact on Avian Species" in *Current Pollution Reports* 10(4), 684–709.

11. Kight, C. R., and Swaddle, J. P. (2011). "How and Why Environmental Noise Impacts Animals: An Integrative, Mechanistic Review" in *Ecology Letters* 14(10), 1052–1061.

12. Nemeth, E., and Brumm, H. (2010). "Birds and Anthropogenic Noise: Are Urban Songs Adaptive?" in *The American Naturalist* 176(4), 465–475.

13. Slabbekoorn, H., and Peet, M. (2003). "Ecology: Birds Sing at a Higher Pitch in Urban Noise" in *Nature* 424(6946), 267–267.

14. De Jong, K., Amorim, M. C. P., Fonseca, P. J., Fox, C. J., and Heubel, K. U. (2018). "Noise Can Affect Acoustic Communication and Subsequent Spawning Success in Fish" in *Environmental Pollution* 237, 814–823.

CHAPTER 3

1. Dumyahn, S. L., and Pijanowski, B. C. (2011). Beyond Noise Mitigation: Managing Soundscapes as Common-Pool Resources: in *Landscape Ecology* 26, 1311–1326.

2. Smith, J. W., and Pijanowski, B. C. (2014). Human and Policy Dimensions of Soundscape Ecology" in *Global Environmental Change* 28, 63–74.

3. Hogan, C. M. (1973). "Analysis of Highway Noise" in *Water, Air, and Soil Pollution* 2, 387–392.

4. Jacobsen, L., Baktoft, H., Jepsen, N., Aarestrup, K., Berg, S., and Skov, C. (2014). "Effect of Boat Noise and Angling on Lake Fish Behaviour" in *Journal of Fish Biology* 84(6), 1768–1780.

5. Whitfield, A. K., and Becker, A. (2014). Impacts of Recreational Motorboats on Fishes: A Review" in *Marine Pollution Bulletin* 83(1), 24–31.

6. Mediastika, C. E., B. C. Pijanowski and A. Pijanowski (2025). "Noise and Its Management in ASEAN and Developed Countries" in *Environmental Science and Pollution Research*. https://doi.org/10.1007/s11356-025-36473-6.

7. Wu, J. (2014). "Urban Ecology and Sustainability: The State-of-the-Science and Future Directions" in *Landscape and Urban Planning* 125, 209–221.

8. Frantzeskaki, N. (2019). "Seven Lessons for Planning Nature-Based Solutions in Cities" in *Environmental Science and Policy* 93, 101–111.

CHAPTER 5

1. FAA (1981). The FAR Part 150 Airport Noise Compatibility Planning Program. https://www.faa.gov/sites/faa.gov/files/about/office_org/headquarters_offices/apl/II.B.pdf

2. FWHA (2025). United States Department of Transportation – Federal Highway Administration. Publications and Statistics, https://highways.dot.gov/resources/pubs-stats/publications-statistics. Last accessed May 2, 2025.

3. United States Department of Transportation, Federal Railroad Administration. (2025). https://railroads.dot.gov/railroad-safety/divisions/crossing-safety-and-trespass-prevention/train-horn-rulequiet-zones. Last accessed May 2, 2025.

4. United States Department of Housing and Urban Development. (2025). https://www.hudexchange.info/programs/environmental-review/noise-abatement-and-control/. Last accessed May 2, 2025.

RESOURCES

For a broad and concise source of information about key federal program policy regarding noise, refer to the Federal Interagency Committee on Noise report titled "Guidelines for Considering Noise in Land Use Planning and Control" from 1980. It contains various tables and a rich source of references including studies and reports by agency.

https://www.rosemonteis.us/files/references/federal-interagency-committee-1980.pdf

A useful document to consult for noise issues and potential solutions (although somewhat dated) is the Noise Guidebook published by the US Department of Housing and Urban Development.

https://www.huduser.gov/portal/portal/sites/default/files/pdf/The-Noise-Guidebook.pdf

Noise Pollution Clearinghouse is a rich online resource for communities in the US and Canada looking for sample noise ordinances.

The National Park Service's Grand Canyon Soundscape Program serves as an excellent example of a soundscape planning effort conducted by one of the nation's premier national parks. It contains useful information about measurements, public engagement, special activities that can be sponsored to increase awareness of natural sources and noise, and works with youth on projects.

http://nps.gov/grca/learn/nature/soundscape-program.htm

International, national, state, and a few local noise ordinance resources can be found at Noise Ordinances Central. Here, resources for developing noise ordinances for indoor and environmental noise is summarized here, mostly for the United States and Canada, with many examples that can be useful.

http://noise-ordinances.com

ABOUT US

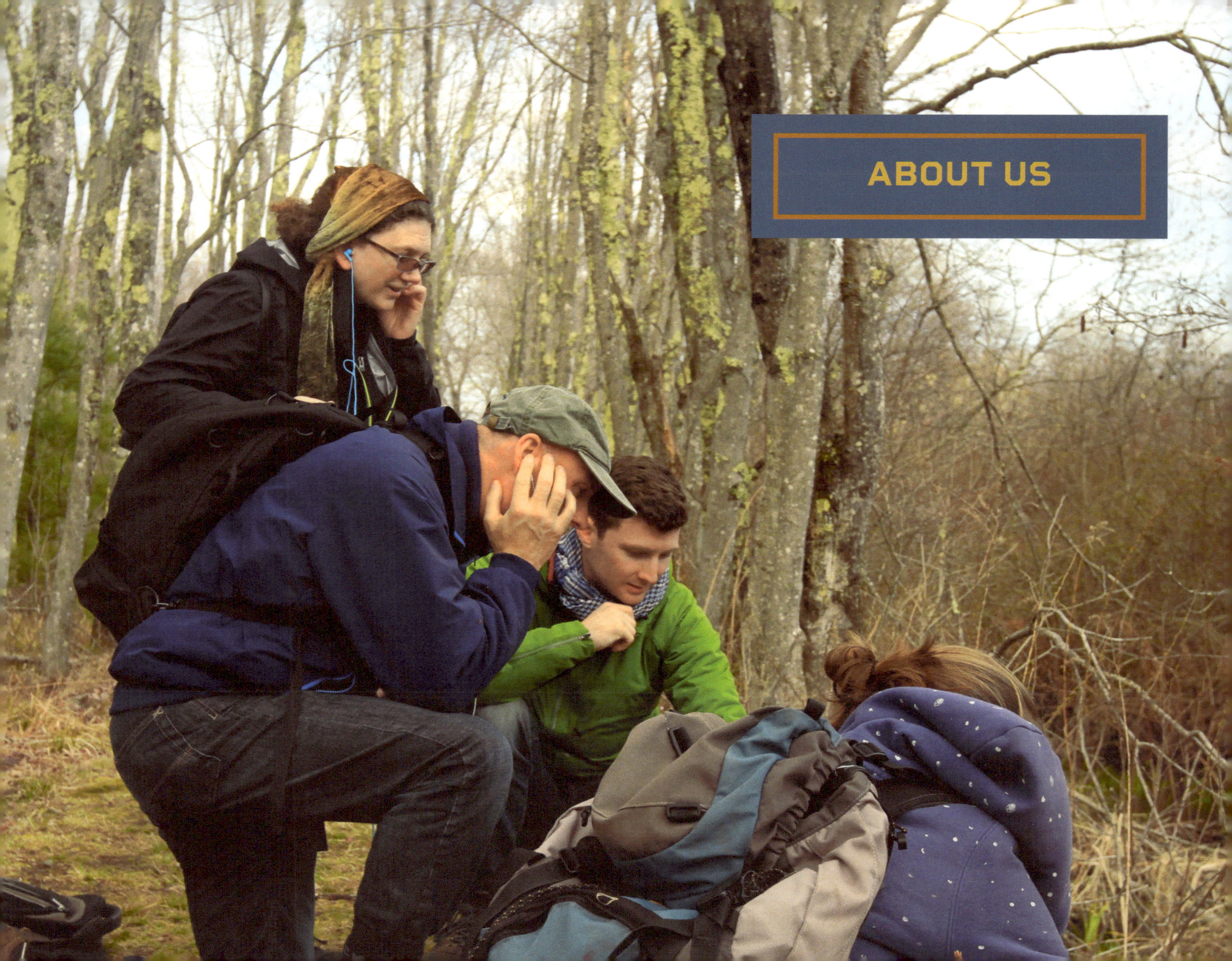

ABOUT THE AUTHOR

Bryan C. Pijanowski

Pijanowski is a professor in the Department of Forestry and Natural Resources at Purdue University. He is also the director of the Center for Global Soundscapes and a recognized leader in soundscape ecology, a field of science and natural resource management that uses sound as a measure of environmental and cultural health. Pijanowski has published more than 180 peer-reviewed papers. He is an American Association for the Advancement of Science Fellow and serves as the executive producer of the giant-screen, big-screen, and domed-theater interactive educational experience *Global Soundscapes! A Mission to Record the Earth*.

ABOUT THE CONTRIBUTORS

Kristen Bellisario

Bellisario received her PhD from the Department of Forestry and Natural Resources, Purdue University. She is a classically trained flutist specializing in modern music performance and analysis techniques. Her PhD research focused on finding unique patterns in soundscapes and how animal communities respond to noise. Some of her recordings include howling coyotes, bats, and many grassland birds. She is currently on the faculty at the John Martinson Honors College at Purdue University.

Javier Lenzi

Lenzi received his PhD from the Department of Forestry and Natural Resources and Ecological Sciences and Engineering Graduate Program at Purdue University. His area of specialization is the effect of human activities on coastal bird breeding biology and in particular on the physiological and behavioral impacts of plastics of generalist-feeding gulls. He has worked in the United States and Uruguay.

Dan Walker

Walker is a community planning extension specialist for the Department of Forestry and Natural Resources at Purdue University. He collaborates with Purdue Extension staff, community leaders, stakeholders, and interest groups within the Great Lakes Region through programs that combine research-based tools with community planning processes to help determine and achieve the public interest. Walker is a member of the AICP and holds a master's degree in urban and regional planning from Ball State University and a bachelor's degree from Monmouth College (IL). He also holds a secondary educator's license in the State of Illinois

David Savage

Savage received his PhD in the Department of Forestry and Natural Resources and the Ecological Sciences and Engineering Graduate Program at Purdue University. His research focused on the applications of acoustic tools to monitor biodiversity in agricultural ecosystems, especially in premium crops such as vineyards and coffee. He has worked on studies throughout the South Pacific, the US, and Colombia. He is currently on staff with the Indiana Utility Regulatory Commission in Indianapolis.

Kara Salazar

Salazar is an assistant program leader for community development and a sustainable communities extension specialist for Purdue University's Department of Forestry and Natural Resources. She holds a PhD in Forestry and Natural Resources, with a focus on natural resources social science. Working with multidisciplinary teams, she develops programs, products, and resources to support community planning and sustainable development strategies in Indiana communities. Her areas of interest include placemaking and enhancing public spaces, lawn and landscaping conservation practices, community development, and natural resources management. Salazar is a certified planner (AICP), a climate change professional (CC-P), and a Professional Community and Economic Developer (PCED) with credentials from LEED AP Neighborhood Development, the National Green Infrastructure Certification Program (NGICP), and the National Charrette Institute

Acknowledgments: We would like to acknowledge the contributions of data collections from Taylor Broadhead, Benjamin Gottesman, and Jack VanSchaik, and design support from Dawn Oliver and Telaina Minnicus.

www.ingramcontent.com/pod-product-compliance
Lightning Source LLC
Chambersburg PA
CBHW041300210326

41599CB00006B/253

9 781626 712034